Moving Plates

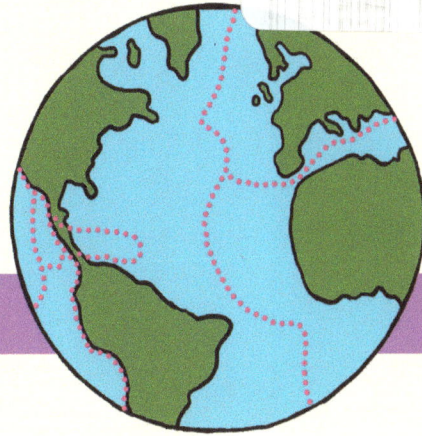

Illustrations: Janet Moneymaker
Design/Editing: Marjie Bassler

Moving Plates
ISBN 978-1-950415-34-2

Published by Gravitas Publications Inc.
Imprint: Real Science-4-Kids
www.gravitaspublications.com
www.realscience4kids.com

Have you ever been
in an earthquake?

The Earth is
quaking!

Have you ever seen

a volcano erupting?

Why does Earth have volcanoes that spout and earthquakes that shake?

Maybe Earth is playing games?

I do not think so.

Earth is made of different parts, or **layers**.

Earth is made of different layers.

The outer layer is the **crust**.

The middle layer is the **mantle.**

The innermost layer is the **core**.

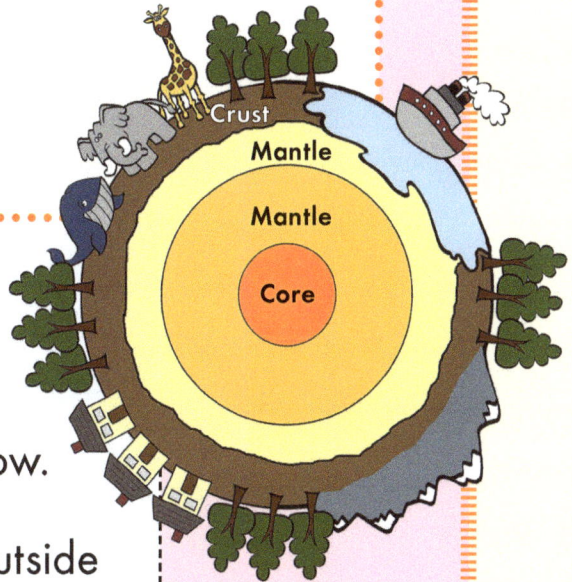

The **crust** is solid.

The **mantle** is solid on top and soft like peanut butter below.

The **core** is soft metal on the outside and solid metal in the middle.

The **crust** and upper part of the **mantle** are solid. These layers are broken into huge pieces called **plates**.

I put pizza crust on a plate!

Silly mouse!

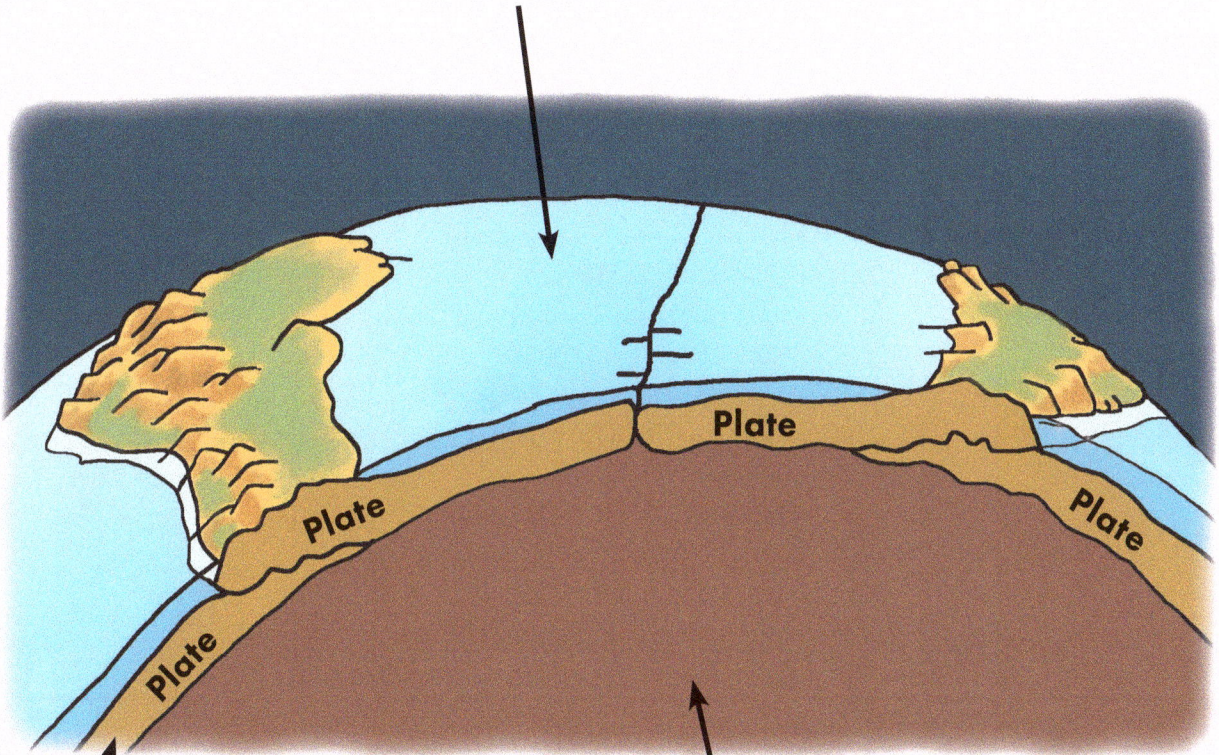

Surface of Earth

Plate

Plate

Plate

Plate

Crust and upper mantle

Lower mantle

These plates float on the lower part of the mantle, which is soft like peanut butter. This soft material is called **magma**. The plates are carried along by the magma as it moves.

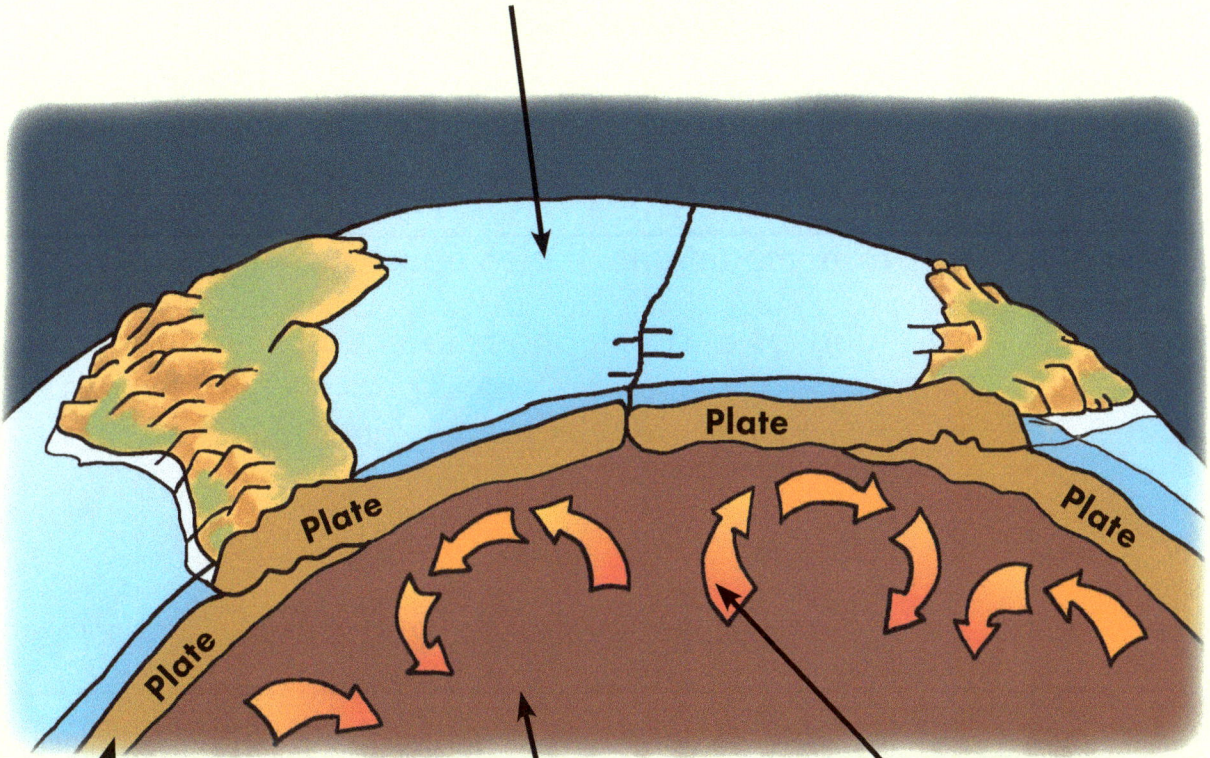

Surface of Earth

Plate

Plate

Plate

Plate

Crust and
upper mantle
(solid)

Lower mantle
(soft, magma)

Magma
moves

Some plates overlap. One plate pushes downward underneath another plate. This can cause the plate on top to crumple and form mountains. It also causes earthquakes and volcanoes to occur.

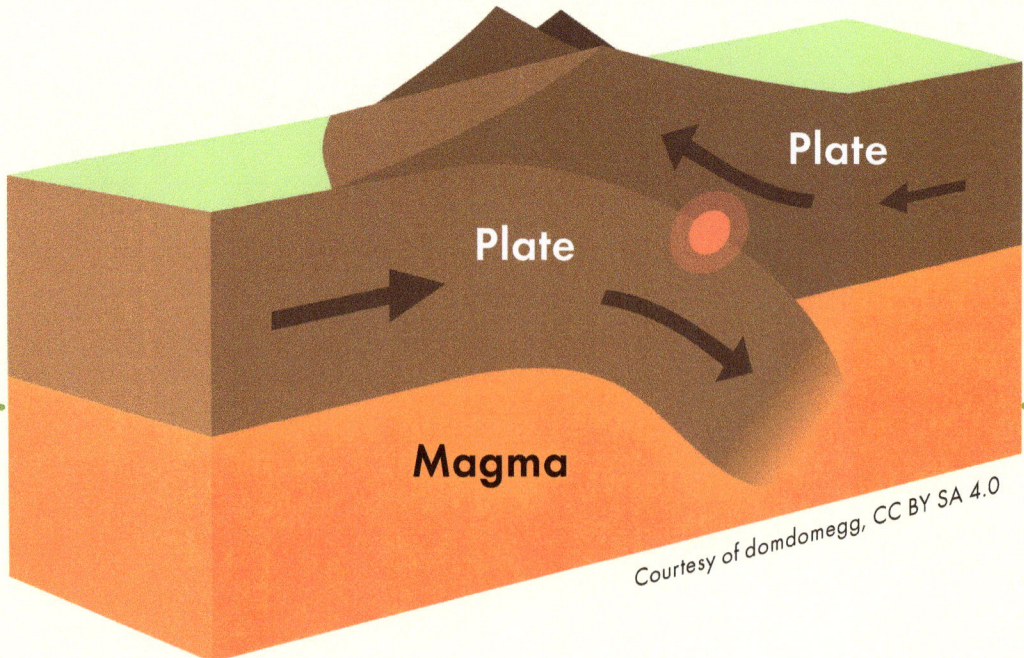

Plate

Plate

Magma

Courtesy of domdomegg, CC BY SA 4.0

Plates can slide next to each other in opposite directions. The grinding of one plate against another can cause earthquakes.

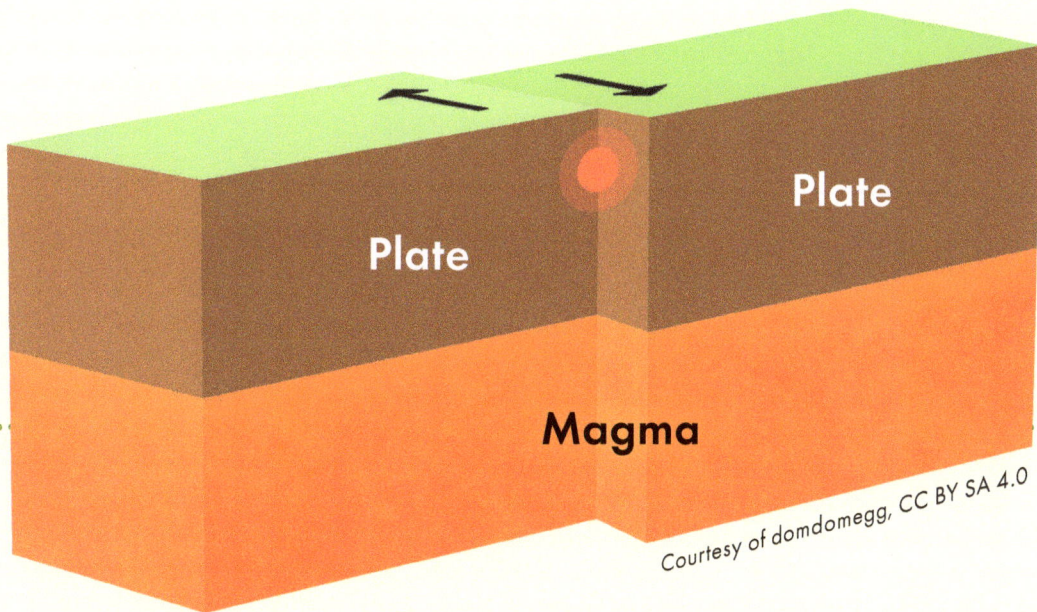

Plate

Plate

Magma

Two plates can move away from each other. This causes magma to move up between them. The magma cools and hardens to form new crust.

Volcanoes and earthquakes take place in this area, which is often under the ocean.

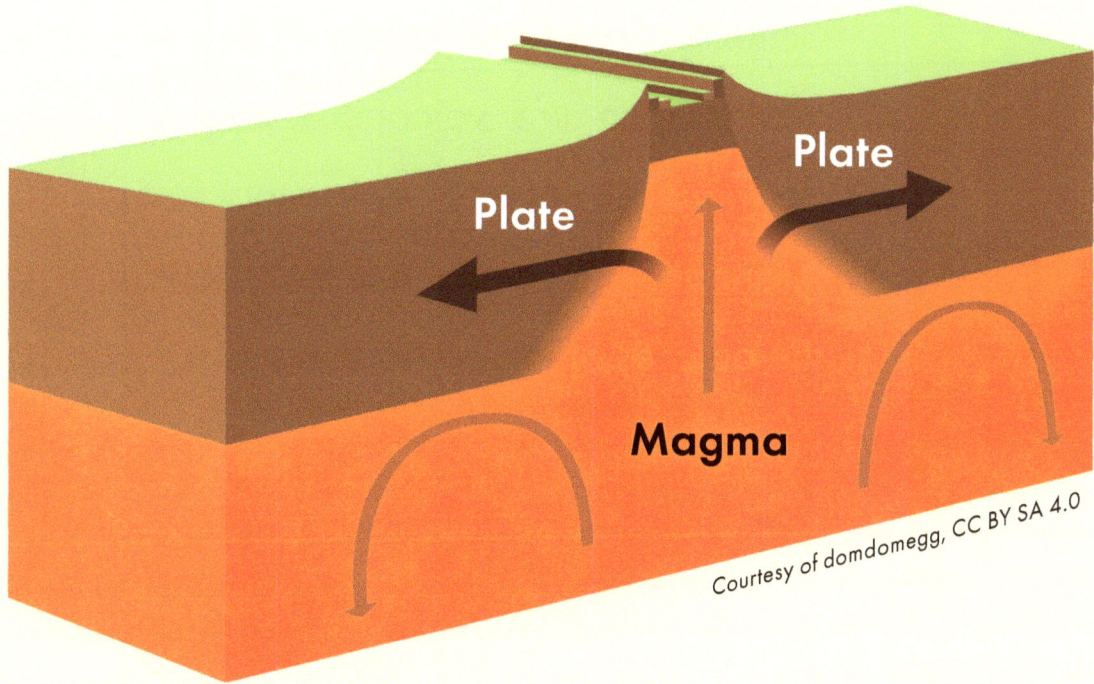

Plate

Plate

Magma

Scientists look for the places where mountains, volcanoes, and earthquakes occur. They use this information to draw maps that show the outlines of the plates that make up Earth.

Those plates are huge!

Earth is huge!

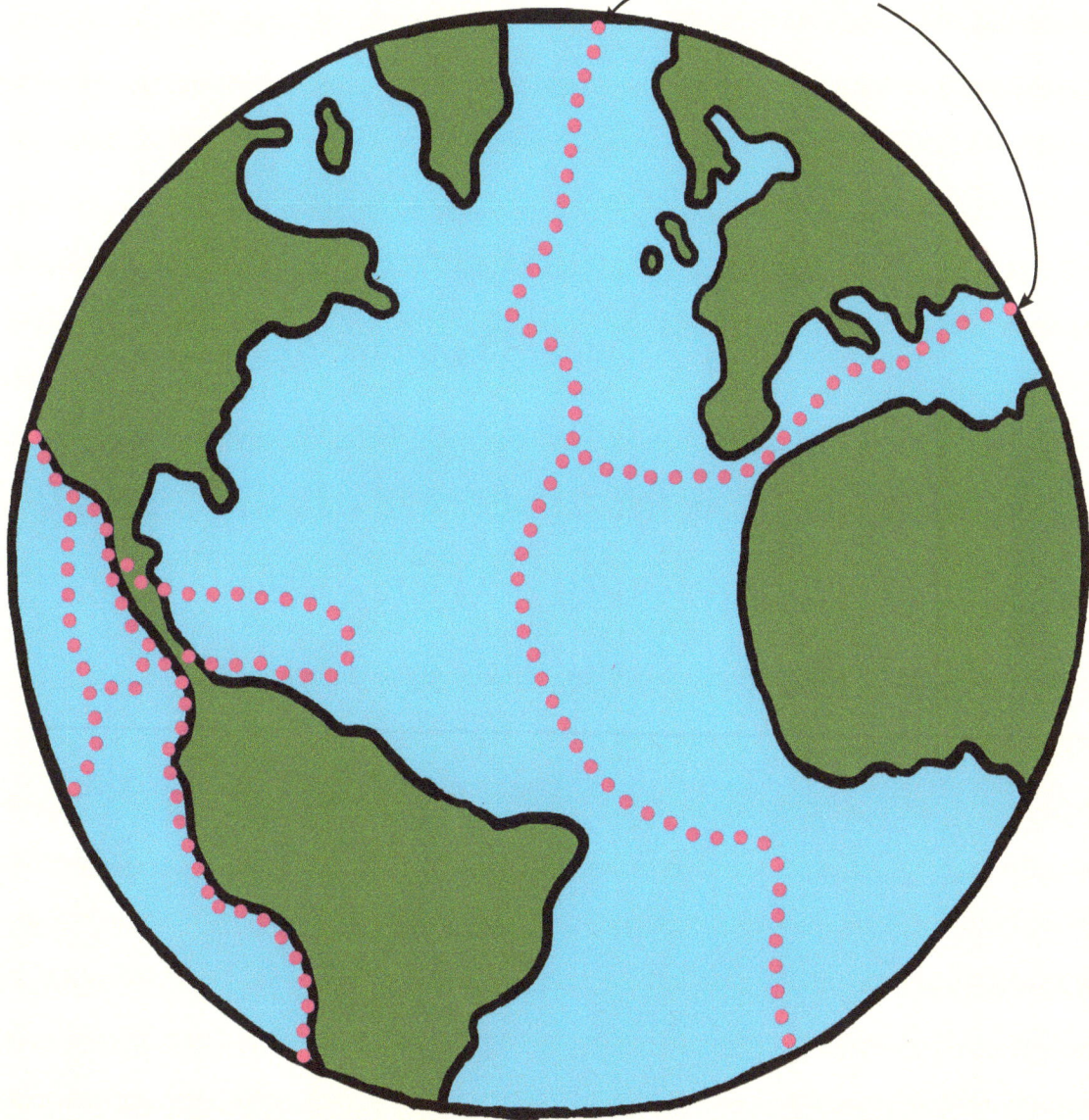

Edges of plates

Learning about Earth's plates and
their movements helps us understand
how mountains are formed and why
earthquakes and volcanoes occur.

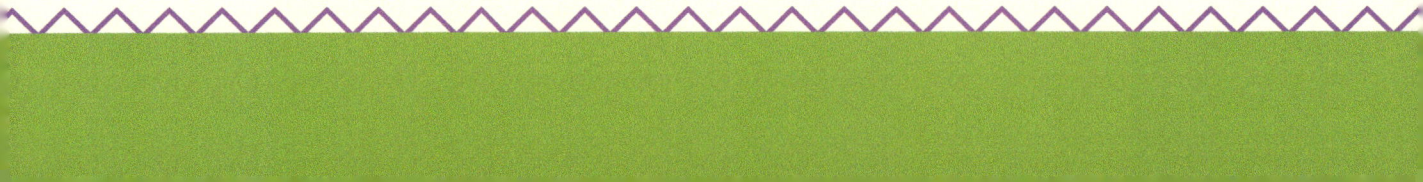

How to say science words

core   (KAWR)

crust   (KRUHST)

earthquake   (ERTH-kwayk)

erupt   (i-RUHPT)

layer   (LAY-uhr)

magma   (MAG-muh)

mantle   (MAN-tuhl)

mountain   (MOWN-tuhn)

plate   (PLAYT)

science   (SIY-ens)

volcano   (vahl-KAY-noh)

What questions do you have about MOVING PLATES?

Learn More Real Science!

Complete science curricula from Real Science-4-Kids

Focus On Series

Unit study for elementary and middle school levels

Chemistry
Biology
Physics
Geology
Astronomy

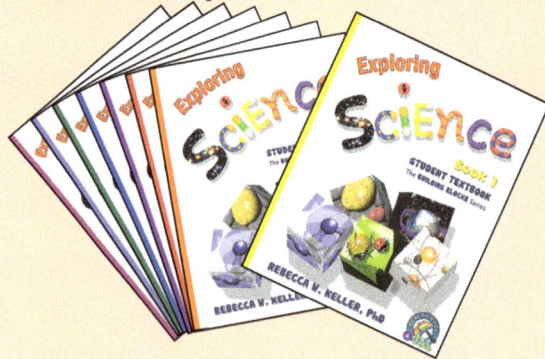

Exploring Science Series

Graded series for levels K–8. Each book contains 4 chapters of:

Chemistry
Biology
Physics
Geology
Astronomy

www.ingramcontent.com/pod-product-compliance
Lightning Source LLC
Chambersburg PA
CBHW040150200326
41520CB00028B/7562